GRIEVING
with
Purpose

Pain to Purpose

JoAnn Sutherland

Trilogy Christian Publishers
A Wholly Owned Subsidary of Trinity Broadcasting Network
2442 Michelle Drive
Tustin, CA 92780

For information, address Trilogy Christian Publishing
Rights Department, 2442 Michelle Drive, Tustin, Ca 92780.
Trilogy Christian Publishing/ TBN and colophon are trademarks of Trinity Broadcasting Network.
For information about special discounts for bulk purchases, please contact Trilogy Christian Publishing.
Manufactured in the United States of America

10 9 8 7 6 5 4 3 2 1
Library of Congress Cataloging-in-Publication Data is available.
ISBN: 979-8-88738-627-0
ISBN: 979-8-88738-628-7

CONTENTS

DEDICATION

I dedicate this book to Flavia Antoinette, my 36-year-old daughter who graduated to heaven. She was God's best gift and allowed me to enjoy her for years. She was full of life, Jesus, and a heart to please Father God all the years of her life. I could not have had a more encouraging loving child. I taught her things but also learned much from her dedicated life to Jesus. She was and is a light that still shines through her two sons, my grandsons Connor and Preston here on earth.

ACKNOWLEDGMENT

I want to thank my husband Rocky, who is a great support to me. He encourages me and helps me see that there is always a progression in God to bring something good out of everything we go through. His love and hugs pulled me through. Praying with him gives me joy. I know when the two of us pray in agreement, it binds us together and brings us closer. After Jesus, he is my rock.

Thank you to my sisters in Christ, Sally, and Amber. Who kept me covered in prayer throughout these last five years. I always knew I had backup which gave me strength.

Abigail, for her taking the time to talk with me and help me get my thoughts together when writing.

My newfound friend and sister, Gloria, for sending me a business card for prayer and then for returning my call with prayer. You gave me the I needed at the right time.

GRIEVING WITH PURPOSE/ PAIN TO PURPOSE

PREFACE

Let me be open and share a little bit of my story. I was a single mom raising and homeschooling a beautiful baby girl. She was amazing, smart, loved the Lord, graduated college, got married, had two wonderful boys, bought a home in Texas, lived there for over five years then moved closer to me and my new husband, Rocky in Idaho. Found out baby number three was on the way. Then, a pulmonary embolism took them both from this earth.

Devastation is not the right word. There are no *right* words to express the screams and yells coming out of my soul. If it wasn't for God's sweet Spirit I would not have survived this, plus the knowing I would see them again.

Here is what I hope helps through your journey of grief. It's not easy and no two journeys look the same. Grief comes in many forms. It could be in the form of death, separation, or even in the abandonment of a parent. Even a four-legged companion who greeted you faithfully after work.

It's been four years and six months, life has started to shine again along with great expectations of a grand family reunion in heaven. Until then, I will use my pain for a purpose-

God's purpose. Join me in doing the same.

Your fellow grieving, healing friend,

JoAnn

CHAPTER 1

SHOCK

Isaiah 46:4b - I will care for you; I will carry you.

**Shock: a sudden upset or surprising blow;
a heightened heart rate, a change in your
breathing pattern**

I got the call between 2:00 am and 3:00 am, can't remember the exact time because as a parent hearing that your child is in the hospital and not doing well, you run like mad fighting every emotion trying to come out at once. prayed, bargaining with God, begging God, pleading, praying, trying to set your head to see a bright future, a sweet future, a happy future. After all , she's pregnant and another grandchild is such a wonderful gift that brings excitement and joy....

Shock- What do you mean she didn't make it?! I'm her mother, let me see her, my God can raise the dead! She's not checking out yet, just give me a minute! Okay give me five minutes; I need another fifteen minutes! It's not over yet! Where are those loud, overbearing screams and groanings coming from? How dare you tell me she's not here, of course she is! I see her!

Did light leave my soul? Did darkness come to take me? Did I do something wrong? Take me please, take me! She's a better Christian than me! The world needs her more than it needs me.

Yes, I'm sure there are many descriptions, but here on this day, I sat numb, waiting for life to pass me by. *Get on with it! Come on life, finish it. You took your best shot and didn't miss.*

You, the reader, may understand this feeling.

Have you taken the time, yes, the hours, days, even months to voice, journal, or record your shock? No? Can't find the words? You don't *feel* it will help? It will. Trust me. Giving yourself a moment to express the aches and pains of your heart through the writing process. Write yourself, God, or your loved one a note or letter. Share your anger, hurt, emptiness, sadness, and yes love for them. Put words to your voice to express the absence you feel from their hugs, smiles, and laughter. Sit down now with the One who already knows how you feel. Father God saw his only son die. Do you think His pain was any less? It may not have shocked Him, but it shocked us when our daughter passed on; a life cut short, she was only 36.

How about you? A child, best friend , sibling, parent, teacher, pet? An unexpected or expected loss takes time to walk through. The shock of it is, well, shocking. We are in no hurry right now.

Let's take a moment to walk through the shock with our Father and Jesus. The Holy Spirit, the comforter will sit with his arms around you. Go ahead, let him embrace you.

1 Peter 5:7

Cast all your worries and anxiety on Me;
for I care for you.

CHAPTER 2
DENIAL

Revelation 21:4 He will wipe away every tear from your eyes and there will be no more death or sorrow or pain. All these things are gone forever.

Denial: refusing to accept something as true, challenging the facts

Why is it when you look at someone who has left earth, it looks like they're still breathing? Is it because your breathing is slower? Maybe it's because you're refusing to blink in anticipation of movement? *Can we check for a pulse again? Let's just sit and wait, maybe God's just testing our faith.* So I wait and cry, trying to hold on to the idea that *it could happen, right? My tears could change from sorrow to joy today. Any moment now.* Well

Today is not that day, *but I know it's coming.*

We can't deny that life is short, unfortunately, shorter for others. We also can't deny that the Father is always with us. Let's take some time and walk through what we can't deny;

- His love

- Our love for the person or pet that is no longer with us

- Father knows our every thought, and every feeling, and knows our hearts better than we do or anyone else does

- Our feelings, pain or loss

- We are never alone

- We are never separated from God's love

> *And I am convinced that nothing can ever separate us from God's love. Neither death nor life, neither angels nor demons, nor our fears for today nor our worries about tomorrow—not even the powers of hell can separate us from God's love. No power in the sky above or in the earth below—indeed, nothing in all creation will ever be able to separate us from the love of God that is revealed in Christ Jesus our Lord.*

Rom. 8:38-39 (NLT)

You're right. It feels lonely right now, but the truth is God will never leave or abandon you. "So be strong and courageous! Do not be afraid and do not panic before them. For the Lord, your God will personally go ahead of you. He will neither fail you nor abandon you" (Deut. 31:6 NLT). Reach your arms up to Him and let Him rain down His presence, peace, love, and comfort.

I'm not asking you to deny the pain, but share it with

the Father. He would not ask you or me to deny our pain, but to share it with Him. We share our pain with Him and He shares His love, life and strength with us. Go ahead, spend a little time and share your sorrow, tears and groanings with your Father and Jesus, the author and finisher of your faith. He can handle it. Let him make an exchange with you.

Isaiah 25:8

He will swallow up death forever! The Sovereign Lord will wipe away all tears. He will remove forever all insults and mockery against his land and people. The Lord has spoken!

CHAPTER 3

ANGER

Ephesians 4:26 In your anger, do not sin: do not let the sun go down while you are still angry.

Anger: displeasure, fury, outrage or a strong feeling of hostility.

Can I hit the wall with my fist? I'll kick it instead! God, can you feel my anger? Breathing is hard right now! I want to scream! Will they kick me out of the hospital away from my daughter's body? I won't go!

Okay, so hours have passed, and they say it's time to go. *Where do you go? Home? Are you kidding me?* I'm angry, I'm seeing red, my fists are tight. *Do not touch me! Don't tell me what to do! I'll go when I'm ready. What is happening?*

Can you relate? Still feeling it? Did I hit a sore spot?

Believe me, I understand. Anger is a good and/or bad emotion/feeling. Good because you *are* feeling something.. It can make the sweetest person *feel* hostile. The enemy wants us to go with that! Just let your anger

21

explode, where Jesus experienced anger and showed it (turning over a table or two, with no cursing, hitting or lingering).

Proverbs 29:11 A fool gives full vent to his temper but a wise man holds it back.

Jas 1:20 Human anger does not produce the righteousness God desires.

So getting mad, and angry is good! Letting it control us? Not good. Get it out. Jesus did.

John 2: 14-17 (NLT) In the temple He (Jesus) saw merchants selling cattle. Sheepand doves for sacrifices; he also saw dealers at tables exchanging foreign money. Jesus made a whip from some ropes and chased them all out of the temple. He drove out the sheep and cattle , scattered the money changers' coins over the floor and turned the tables over. Then, going over to the people who sold doves, He told them,"Get these things out of here. Stop turning my Father's house into a marketplace!"

Let it out and then move past it. Anger festering will only grow! Scream , cry, talk, shout, throw something (preferably away from everyone)! Walk, run, hit a punching bag, or do something you can to release the anger. And by the way, it may take more than one time.

Let's make time to get through some anger,you are not alone. Remember that!

- Walk_____

- Talk_____
- Run_____
- Cry_____
- Boxing_____
- Scream_____
- Shout _____
- Throw something_____

Then repeat.However, many times are needed.

CHAPTER 4

BARGAINING

Genesis 32: 24-26 (NKJ)Then Jacob was left alone a Man wrestled with him until the breaking of the day. Now when he saw that He did not prevail against him, He touched the socket of his hip, and the socket of Jacob's hip was out of joint as He wrestled with him. And He said, "Let Me go, for the day breaks." But he (Jacob) said, "I will not let You go unless You bless me!"

Bargaining: making an agreement, a pact, a pledge. Negotiating a deal.

Take me! Not her! Not my only baby! Not the one you gave me! Not the joy of my life! My heart can't take it! I can't breathe! I'll give you anything you want! I'll sleep four hours a day and pray twenty! I'll never do another mean thing! I'll never criticize! I'll watch my mouth and help others more!

Making negotiations comes along in our mind like it will make everything better. Change what is happening, perhaps.. Cancel out divorce papers, raise our child from the dead, breathe life into situations that are failing. We start with small agreements or pledges and then keep increasing them. Until we offer up our very lives for the lives of others.

1 Kings 19:4b And he prayed that he might die, and said: It is enough now, Lord take my life for I am no better than my fathers!" (NKJ) *Please take me! I'll take their place without a fight or fuss!* Later, as you go back to rethink these thoughts. You will see the bargaining phase, *(if you linger for quite a while),* as the tormentor who comes to take you out and those around you down. Matthew 16:22-23 (RSV) And Peter took Him (Jesus) and began to rebuke Him, saying, "God forbid, Lord! This shall never happen to you." But He (Jesus) turned to Peter, "Get behind me satan! You are a hindrance to Me; for you are not on the side of God, but of men." Staying in the bargaining mindset too long sets you up for discouragement and disappointment. If you just lean on the Father and trust He is there to help you walk through it. The bargaining phase won't last long!

I'm encouraged and hope you will be too! Though it is very aggressive and unpleasant, it is short-lived.. God listens to your bargaining, He understands. It doesn't upset Him at all. He's heard it many times. (2 Samuel 12:16) David wanted to bargain for His son's life, wouldn't we all? Of course we would or for a spouse or a parent. Once we attach our love, our life to someone, letting go is a difficult choice: grieving, negotiating and bargaining with

God is a sign of fear. Fear of trusting the only One who knows the beginning to the end. Revelation 1:8a "I am the Alpha and Omega," says the Lord God. (Revelation 1:8a RSV)

During the bargaining phase, the one thing that can and will help is getting quiet. After your thoughts have run the gamut, get quiet. Yes, tears will probably be flowing because *reality,* as the world knows it, is settling in. Be still for a moment. Psalms 46:10 (NKJ) "Be still and know that I am God the Father, you God comfort you."

Matthew 5:4 (NKJ) "God blesses those who mourn and comforts them (bestows comfort on you)."

Let Father God show you His compassion, His understanding, His love).

Lamentation 3:32 (DBY) but if he has caused grief, he will have compassion according to the multitude of his loving-kindness.

Quietness can be a friend or an enemy. Depending on what we focus our thoughts on. I often remember my daughter with joy, laughter, and full of happy times. And yes, even with a, huh? And what were you thinking? She was quite the child. Full of adventure. Sometimes my thoughts go to anger or disappointment *because she's gone. Isn't that all part of life?* I've allowed silence, and quietness with the Lord to bring back my memories with her, sometimes with tears but always with the knowledge that I'll see her again. Quietness is my friend. I don't sit back in anger at her or God because I can see brightness ahead. So sit my fellow griever. Yes, lament, cry and then, be still. So together you and your Father can see the bright future He has for you.

Now, I know and understand we have walked through a lot of tears, screams, anger, shock, and bargaining, but stay with me. Once we get through the grieving, even as we are going through it, Father God is revealing more and awakening our purpose in it, showing and reminding us we are still here, we still have a purpose! More on that soon!

So journal now. Write and share with the Father what you are trusting Him with and trusting Him to do.

Psalm 31:14 (NLT)

But I trust in thee, oh Lord, I say, "Thou (You) are my God."

He's listening.

CHAPTER 5

DEPRESSION

Psalm 6:6 -7a (NLT) I am worn out from sobbing. All night I flood my bed with weeping, drenching it with my tears. My vision is blurred by grief; my eyes are worn out.

Depression: sad, discouraged, oppressed, downcast, sorrowful, dispirited

Breathing seems difficult. I think I slept some, I woke up startled, heart pounding. I feel like I went for a run! So tired,so, so tired. Is it cold in here? Where are my shoes? Why do I need shoes? I should just lay here, why get up? Can't cry anymore, tears? They're gone. I want to cry though, I can't scream, but I want to. I'm numb. What do I feel? Nothing! Why am I cold?! Maybe I should get up, but why? If I lay here, maybe God will have mercy on me and take me now.

Sound familiar? Feel familiar?

Are you thinking about how those around you would be better off without you? Because you are just toodepressed, numb, tired? Are you feeling useless? Worthless or in the way? Not eating? Eating too much? Depression sucks! I mean literally, it sucks the life right out of you! *Where did I put my keys? Did I take my vitamins? Did I brush my teeth?* Forgetfulness, yep. Depression can make you feel like you're out of control, emotionally and mentally. It also affects you physically. Depression can cause joint pain, digestive problems, headaches, and can weaken your immune system.Depression bites! You and I cannot deal with depression on our own. We don't have to!

Isaiah 41:10 Fear not, for I am with you, Father God says, He will strengthen us, help us. He will hold us with His right hand.

Isaiah 42:16 He will turn the darkness into light, and make the rough places smooth.

Father knows things are rough right now. Hard, out of focus, heavy, and just downright overwhelming. He understands a crushed spirit, a heartbroken parent, a left-behind spouse, a close sibling, or a forty-year friendship. And sometimes it's that constant companion of a pet you've had for ten years. For a single person, that four- legged greeting every day at the door had special meaning. We have help, we have promises , we have Jesus who understands loss and a heavenly Father who gets it and experienced it also.

Psalm 34: 17-18 (NLT) The Lord hears you when you call to Him for help. He rescues them from all their troubles. The Lord is close to the brokenhearted He rescues those whose spirits are crushed. *(overwhelmed or disappointed)*

Matthew 5:4 (NIV) Blessed are those who mourn *(feel deep sorrow)*, for they will be comforted.

You see, in Matthew 5 (NIV) Father God says He will bless you, *(consecrate you, anoint you),* and comfort you, *(help you, support you, ease the grief or pain).*

As you lean on Him, sit in His arms, His presence. Whether it's five minutes or two hours, He will take, carry, remove and hold all your hurt and pain. But He won't snatch it from you, *you must* give Him your hurt, your sorrow, your mourning. Don't hold on to it. Hand it over! Let go of it! Come on, you *can* do it.

It won't be easy. Nope, and how many times will it take before you can completely let your depression go? Put it in His hands. Two, three, ten times? That doesn't matter. How many times do you try to handle it yourself before it will try to drag you down to the deep end of the pool.

This stage can last the longest and be the most draining. If you try to walk through it alone, I might understand . You don't have to. None of us do. So I didn't . Whether you have a friend or family member to talk to lean on or just sit with, remember you also always have your heavenly Father.

Psalm 9:9 (NIV) The Lord is a refuge for the oppressed *(depressed)*, a stronghold in times of trouble.

We have to allow the Lord to shelter us and protect us. Be our refuge and our safe place. His arms are our stronghold, our fortress in times of unrest, difficulty or mental fatigue.

Psalm 73:26 (NET) My flesh and my heart may grow weak, but God always protects my heart and He is the source of my strength.

You are *not* alone! Let me say that again. You are *not* alone! Never ever alone.

Let your Father God be your strength, your solid source of strength. Let Him give you the power to go on, to endure.He is so much bigger and stronger than we could ever imagine. After all, He is Omnipotent. All powerfulOmniscient all-knowing . Omnipresent, always with us.

He will be your strength. Your light at the end of the tunnel. So keep moving. Get up and take a bath. Go for a walk. Make a phone call. Eat something. He will steady you with His right hand. Psalm 73:23 (NIV)

He will make you strong.

Why am I spending so much time with you on this? Why all the scriptures?

Because this stage of grief, the sadness, the longing, and the separation of intimacy with the ones we love can provide an easy target for the enemy to hit. He will do his best to keep you isolated and stoic. Indifferent to life. Your life, your family's lives, your Christian life. Now captivated by sorrow, the enemy can keep you here longer and the more damage he can do to you physically and mentally, and to those around you. Depression is a silent place. We feel we have no voice, no emotions, no strength. We want to talk to our loved ones one more time. Hear their laughter again. Look into their smiling eyes and say 'I love you' one more time as we hug them. Resting in depression, yes sitting in that state too long can literally take you out of commission.

I'm not saying there is a time limit on any stage of grieving. Hear me now! Take your time. You do you.

But, allowing depression to run amok for a lengthy period of time causes a persistent, constant feeling of loss, and sadness.

That is *not* God's best for anyone. Jesus came to give us life more abundantly. (John 10:10 NKJ)

You see Father God knows those feelings you are experiencing and understands the pain, the anguish, the loss, the separation. He was separated from His son. No contact. No laughter. No talking. Silence.

He feels the same loss from you when you turn away from Him or don't let Him comfort you. His love is so great, so longing for closeness with you. To hear *your* laughter, to wipe *your* tears, to talk to you and say to you, 'I love you, let Me hold you, talk to Me.'

We have to remember, we are still here! Still breathing. Still able to fulfill our purpose. We have been commissioned to go and do. Yes, we are called to *go* and *do*.

Mark 16:15 (NLT) And He told them, "go into all the

world and share the *good news* with everyone."

2 Timothy 4:2 (NIV) "Go teach the Word; be prepared in season and out of season; correct, rebuke, and encourage, with great patience and careful instruction."

I must admit, as I *went* and *did*, I started to feel alive. Breath again. Connected to life again! Wow!

Sounds scary? Too hard to do? *Go* and *do* it alone? Never! Don't you worry.

Joshua 1:9 (NLT) "Don't be afraid or discouraged God is with you wherever you go."

What is life?

A gift from God. Each of us is given just so much of it. Then it's gone, like a morning fog.

James 4:14b (NLT) Your life is like the morning fog, it's there a little while and then it's gone.

While Jesus was on earth in His human suit, He experienced love, laughter, happiness, sorrow and loss.

And because of His gift of life for us, look what we have! Eternal life. One day we will get to enjoy life unendingly . Without loss.

Jesus is the example, we must *go*, yes, let our life *go* just as He did. Live our lives with purpose. As we do this we will find our future and purpose. As we allow Jesus to live through us, as we help others along the way.

Matthew 16:25 (NIV) "For whoever wants to save their life will lose it, whoever loses their life for Me will find it."

Luke 17:33 (NLT) "If you *cling* to your life you will lose it, and if you let your life go you will save it."

If you cling hold tightly to or clutch, death grip your life, you'll lose it.

Sitting, laying around, and waiting in a depressed state is losing *your* life. It's not *going* and *doing*.

Do you know how you can really do life? By living it! Getting up and lifting someone else up with you. Easy? No.

Doable? Yes. A slow start may be all you can do right now. Fantastic! Do it.

So enough of this depressing talk! Time to smile.

Let's start our walk through this valley of lifelessness and wipe our eyes on a tissue or sleeve, blow our nose, splash some water on our faces . Allowing Father God to begin a resurrection; a restoring of life to our soul (*emotions*), our passions, our feelings and our vitality (*activeness*).

How do we become resurrected? We can't do it ourselves. It's through the *power* of our Almighty heavenly Father.

Ephesians 1: 19-20 (NLT) I also pray that you will understand the incredible greatness of God's power for us who believe in Him. This is the same mighty power that raised Christ from the dead.

Ephesians 3:20 (NLT) Now all glory to God, who is

able, through His mighty power at work within us, to accomplish infinitely more than we might ask or think.

Yes, , if you have received Jesus as your Lord and Savior, then you have the Holy Spirit dwelling in you.

You are now primed, loaded and ready for the restoration of the life Father God offers. So just ask your heavenly Father to resurrect you. His purposes, His plans and His future for your life.

Jeremiah 29:11 (NLT) "For I know the thoughts I have for you," says the Lord. "They are plans for good and not for disaster, to give you a future and a hope."

He will help you every step of the way as you *go* and *do*.

Psalm 121:1-2 (NIV) "Where does my help come from? My help comes from the Lord, the Maker of heaven and earth."

Now go! But first, bask in and accept His help and power. The Holy Spirit is ready to release His power in your life to help you.

What is the Father wanting to resurrect in your life?

What purpose, plan or activity? So it's time to pray and listen.

And make some notes.

Acts 1:8a (NLT)

"But you will receive power when the Holy Spirit comes upon you."

_____

_____

_____

_____

_____

_____

_____

_____

_____

_____

_____

_____

_____

_____

_____

CHAPTER 6

ACCEPTANCE

Matthew 26:39b (NIV) "My Father, if it is possible, may this cup be taken from Me. Yet not as I will, but as You will."

Acceptance- welcoming, embracing, accepting, corroborating

I accept my life is different now. Not good or bad, just curving in another direction.

Different.

I accept the fact that when I want to talk to or hug my daughter, I will not be able to, for now.

Understanding that *for now* our absent friend, spouse, parent or pet will not answer back, hug us or talk to us, will take some time to accept. We must embrace our separation and know we can and will continue to remember them, their lives, the joy they gave us, and the great times we had. together. Acceptance is something we 'acquire' or come into over time . To walk in, procure, and achieve acceptance will take some effort.

Do you remember that first time you wanted to call that friend, spouse, child or parent to share some good news, or to ask for some help with a problem? Bummer, right?

Or you rush home to watch your favorite TV show and reach for the phone to call your mom, but not this time. You open the front door waiting for Daisy, your dachshund to greet you, but not this time. You do a shout-out for your spouse to check the back door and make sure it's locked, only now that's your responsibility. Or I hope dinner is ready when I get home, I'm hungry. Getting home only to recall, oops, I guess I'm having a can of soup.

Acceptance can lead us right back to the depressing train again...Choo Choo! Let me encourage you here. That train doesn't become any more fun the second time around.

Effort.

Embracing and welcoming some 'new' or 'renewed' things in our lives can take work, is laborious, and downright hard.

Psalm 23:3-4 (NLT) He *renews* my strength, He guides me along the right path. Even when I walk through the darkest valley, I will not be afraid, for you are close beside me.

It will take courage to walk the path of acceptance, accepting Life has mountains and valleys. Some days. acceptance will require some extra inner strength to move forward. Some nights may have you starting new bedtime rituals. Some dinners will have you embracing *new* or 'renewed' talents like cooking. Or ordering out! Can we say 'hallelujah',

Praise the Lord for takeout pizza or Chinese ! Can I get an 'amen' for tacos and burritos?

Acceptance doesn't mean you have forgotten them or left them behind. Or that you think about them only on occasion from the rearview mirror of life. It is just a way of thriving and living with the vibrant memories that bring us joy, tears, smiles, and the hope we will see them again.

They make their 'appearance' in our thoughts often. We have *accepted* this as enough for now, and begin to laugh again and live again. Especially when we think of our times with them.

I find myself at times laughing out loud in the car about something my daughter said or did.

It is a wonderful pain that I don't want to be without, because of the gift of time I had with her.

It is all precious to me and worth the short spurts of sadness to remember her.

So now in the acceptance process, let the Lord give you new strength to keep going. To see each new day as a new opportunity to celebrate His goodness and mercy. Trusting Father God to encourage and help you. To gain new strength to continue on your journey with purpose.

Isaiah 40:31a (NLT) But those who trust the Lord will

find new strength.

With acceptance, you'll find that *hope* starts to raise its pretty little head!

Dreams start to appear. Aspirations, and desires, like flowers, begin to bloom in our hearts.

Winter and dark days make way and lead us to Spring with a new. . Cravings and longings start to bud anew.

One might say, we get a *hankering* to do something!

However long it takes for us to get here is different for everyone. No comparing or criticizing.

God made us all unique individuals. Let's never forget that, but be patient with each other. Let 'hope' grow whether quickly or slowly. Be patient.

Ephesians 4:2 (NLT) "Be patient with each other, making allowance for each other's faults because of your love."

What has acceptance 'sprung' up in you? What new or renewed hankering has started to rise up in you?

Feeling ready to use some of your grief for purpose? I know the feeling. Enjoy.

Chapter 6: Acceptance

Take a moment just to write out a few inspiring thoughts you are having!

_____

_____

_____

_____

_____

_____

_____

_____

_____

_____

_____

_____

_____

_____

CHAPTER 7
HOPE

Ecclesiastes 9:4 (NLT) "Anyone among the living has hope."

Psalm 62:5 (NIV) "Yes, my soul finds rest in God, my hope comes from Him."

Hope- desire, plans, hankering, daydream, anticipate

Some will walk witha stride into *hope*, and others, well they may drag their feet. It's okay either way.

Like a seed, it will grow and produce.Tears have watered the garden of hope for days, weeks, months, and years. Like flowers after a rain. The seeds of purpose have become buds of hope. Don't neglect them. Feed those sprouts of hope with the water of God's word, fertilize them with prayer and encourage them with songs and hymns. Let laughter and words from friends, pastors, and family members be the sunshine.

You are entering the process now!

Now it is time to write a letter.

Tell your heavenly Father how your faith and hope are returning. That you are starting to see the sun (and Son) peeking out from behind the clouds.

1 Corinthians 13: 13 (NLT)

"Three things will last forever- faith, hope, and love, and the greatest of these is love."

_____

_____

_____

_____

_____

_____

_____

_____

_____

_____

_____

_____

_____

_____

CHAPTER 8

THE PROCESS

2 Corinthians 9:8 (ESV) "And God is able to make all grace abound to you, so that having all sufficiency in all things at all times, you may abound in every good work."

The Process- activity, task, to begin or continue a course of action

The process of seeing your grieving and pain turn into purpose is a wonderful and exciting moment.

It is time to write down those dreams, those desires, plans, longings and aspirations.

It's time to move into purpose. Let your hope inspire you. It has grown into a purpose.

Job 14:7 (NLT) "Even a tree has *hope*. If it is cut down, it will sprout again and grow new branches."

You may feel like you've been cut down or maybe just had a pruning experience. But it is time, it is the season to grow new branches. To walk in a new purpose or a renewed purpose, that Father God created for you. Let your grieving motivate you to fulfill your purpose with a vengeance.

Exodus 9:16 (NIV) But I have raised you up for this very *purpose* that I might show 'you' my power and that my Name might be proclaimed in all the earth.

You have a purpose, right here, right now. This is your time to do, fulfill and move into it.

You may say, I don't know what it is. Or ask, if I have a purpose, what is it? How do I find it?

Maybe you are questioning, why would I have a purpose. Didn't my child, spouse, or friend?

I can understand your questioning, your concerns, and trying to understand.

Whether your child, friend or spouse fulfilled their purpose is between them and Father God. Just as you fulfill your purpose is between you and Father God. Father God's intention and/or purpose for creating you has never left you or stopped being. It has always been there. Never to be withdrawn or squashed.

Romans 11:29 (NLT) For God's gifts and His call can never be withdrawn.

(NKJ) For the gifts and the calling of God are irrevocable.

Let it gain some priority and focus in your life from you. Think about this for a moment.

You are here for such a time as this. What recurring

thoughts keep repeating in your mind over the years?

Not sure? Can't think of anything right now? Not to worry. You just ask your heavenly Father. He hasn't forgotten it or you.

Psalm 57:2 (NLT) I cry out to God most t high, to God who will fulfill His purpose for me.

Cry out to God to clarify, reveal and help you fulfill the purpose He created you for. To bring the vision back up in your spirit, so you can do it with a newfound desire, joy and strength. Yes, joy and strength! Sorrow and grieve no more!

Nehemiah 8:10b (NKJ) Do not sorrow, for the joy of the Lord is your strength.

Now that you have walked through the valley and headed back up the mountaintop, God has great plans for you. He always did. Now because of your renewed strength, He can use you to help another sister or brother. Part of your purpose whether you are a homemaker, manager, truck driver, teacher, doctor or waitress, is to care for others and lead them back from their grieving to purpose. You will be able to do this with such care and empathy. Comforting others as you have been comforted.

> *God is our merciful Father and the source of*
> *all comfort. He comforts us in all our troubles*
> *so that we can comfort others. Whenthey are*
> *troubled we will be able to give them the*
> *same comfort God has given us.*
>
> **2 Corinthians 1:3b-4 (NLT)**

Father God through His Spirit has filled you with a new perspective. He gave you such a strong and unrepentant purpose, only for you to do. Only you can fulfill the calling and purpose He has placed upon your life.

Admit it.

Your motivations, intentions, goals and ambitions have been revitalized. Renewed. Brought back to life. You've noticed you look up more now than down. Things that were out of focus have become clearer, calling for your attention. The things of God seem to be calling you, drawing you. Asking you to join in and participate in His purpose for your life. Do you feel it? Sense it? Want it now?

I did and do.

People's lives and stories began to draw me, call me. I found myself focusing more on those around me. I started to desire (*purpose*) and aspire (*purpose*) how I might be of service to those around me. Helping in whatever way I could. Whether it was buying gas for a stranger, praying for a co-worker with medical problems, hugging a mom whose child committed suicide, or giving money to a distraught person who could be homeless. My purpose became more and more like Fathers.

Ephesians 5:1-2 (NLT) Imitate God therefore in everything you do, because you are His dear children.

Live a life filled with love, following the example of Christ.

We have got to come to the understanding that, though we may have purposed to be a teacher, doctor, manager or homemaker,God is multifaceted. We are made in His image. God has *purpose* within a purpose. So we do also.

Genesis 1:27 (NKJ) So God created man in His own image; in the image of God He created him; male and female He created them.

There is so much more to us than we know or realize. That only our heavenly Father can show us.

So let's pause for a moment.

This is a lot to think about.

Purpose, image, going and doing.

Take a moment and think about the people in your life. The ones you've noticed a change in their countenance. The co-worker who once had a bounce in their step who now seems preoccupied. The child that now looks away instead of at you.The criticizing relatives you avoid. The lady behind the counter who stopped smiling.

You are where you are right now because God wants you there. Right now, for such a time as this.

Esther 4:14b (NIV) And who knows but that you have come to your royal position for such a time as this.

Your position, your location, your situation.

Yes, you now stand in your *place* with strength, purpose

and insight others don't have yet.

Take a moment.

Jot down a few names or situations that God brings to mind. Leave space next to them and ask the Holy Spirit what He would like you to do, say or pray for them.

Let Him show you His purpose, His goal for using you in their lives. The end *joy* will be great! Amazing! Worth it. Fulfilling your soul.

I pray that the joy of the Lord and the strength He gives you, motivate you more and more each day.

You are a light. Let your light shine now, no matter how dim it may seem to you. Let Jesus shine bright through you. You came from the Father of lights and will return to Him one day. So before you go, share that light.

Matthew 5: 16 (NKJ) "Let your light so shine before men, that they may see your good works and glorify your Father in heaven."

For such a time as this: *would you have me to do for those who are hurting?*

_____

_____

_____

_____

_____

_____

_____

_____

_____

_____

_____

_____

_____

_____

_____

_____

_____

CHAPTER 9
REMISSION

*Matthew 26: 28 (NLT) "For this is My blood of the
new covenant, which is shed for many for
the remission of sins."*

**Remission- the disappearance of symptoms,
forgiveness, pardoning, absolution.**

I really wanted to address this a little. As we walk through the stages of grief, some of the stages may find us thinking, saying or doing things we may later regret. Whether to ourselves, others or God.

We need to take some time to forgive.

Ourselves.

Ask forgiveness of others and yes Father God. If you are holding on to emotions, or feelings that have you seeing Him as unfair or unjust.

Father is just and fair. Asking Him to forgive you is for you. It frees you from keeping a distance from Him. And from your thinking He's holding a grudge against you. Or that there is a chasm between you and Father, that His love can't cross over. Not true.

Now let's look at
Re- regarding, concerning
Mission- purpose, calling

In *remission*, the signs of anger, frustration, hurt, and depression are covered with the blood of Jesus.

And make way for *re-mission*, leading us back to or regarding our purpose, calling.

His goal for our lives. During the grieving stages, we generally lay aside any purpose or plan that was at the forefront of our minds. Any leading we might have felt, we ignore. Being totally immersed in loss and the side effects of loss.

There's nothing wrong with being distracted for a time or season. But then as we see again and again in the word of God, *we must* return or begin anew to focus on our purpose. We have an assignment here on earth. Until we are escorted into the presence of God, we are called to fulfill it. Losing sight of it for a length of time can be devastating for us and affect those around us.

King David lost focus of his assignment and purpose. (2 Samuel 11:1-4)

If we don't turn back to our purpose eventually after grieving, we can become disillusioned, and distracted. That takes us nowhere or down a path we later regret, pulled into other things the enemy lays out for us.

Similar to Sarah, Abraham's wife. (Genesis 16:1-6)

Your God-made purpose is only for you to fulfill.

Peter lost focus *while* walking on the water! (Matthew 14:28-30)

Keep your eyes on Jesus, He will get you to your purpose and help you finish your race.

Let's look at Peter again. *Not to pick on him!* (Matthew 16:16-23)

Peter starts out strong. Jesus even called him a rock, only to tell him to get behind Him a few sentences later. Peter lost focus of his purpose. The plans and calling Jesus had for him. No worries, he made it back.

We all can lose focus of our purpose or God's plan for our lives. Especially if we encounter a time or times of grieving. Some of us may experience more than one loss in our lives.

I know not a happy thought.

We never walk through a valley of loss or grief alone. Our Father of love is always with us.

Lamentations 3: 32 (NIV) Though He brings grief, He will show compassion, so great is His unfailing love.

He will never let you go through anything alone. He's there even if you don't *feel* Him or want Him to be.

He's working behind the scenes. He's there.

Joshua 1:5b (NIV) So I will be with you; I will never

leave you nor forsake you.

He will help us to remember, refocus and fulfill our purpose.

The residue of loss is becoming like precipitation. Dew on the morning grass. Sunshine after a rainy day.

Let it continue tofeed and water your soul as you grow in purpose.

Journal those last thoughts *now* of concerns and doubts.

Forgive, let go, and let's get to destiny.

Psalm 139:2 (NIV)

*You know when I sit and when I rise; you perceive
my thoughts from afar.*

_____

_____

_____

_____

_____

_____

_____

_____

_____

_____

_____

_____

_____

_____

_____

CHAPTER 10
DESTINY

Romans 8: 29-30 (NIV) And we know that in all things God works for the good of those who love him, who have been called according to His purpose. For those God foreknew He also predestined to be conformed to the image of His Son, that he might be the firstborn among many brothers and sisters.

Ephesians 1:5 (NIV) He predestined us for adoption to sonship through Jesus Christ, in accordance with His pleasure and will.

Destiny:future, predestination, portion

Destination, heaven!

But not quite yet. We need to finish our race. Fulfill our purpose.We were chosen, set aside, and predestined to be here now.

This is our *portion*, our part.

Lamentations 3:24 (NIV) I say to myself, "The Lord is

my portion; therefore I will wait for HIm."

How many others are also predestined and don't know it yet?

Some knew from the first church service they went to that they were His.

Others were nurtured, loved and given all of life's luxuries on a silver platter, but never were served up the truth that they are a child of the King. They never heard they were made with a purpose and plan just for them. Now there are some who never made it to that first church service. Whose family said they were nothing special. Worthless, unexpected, unwanted.

Whether you land in one of these categories or not, you cannot sit in the middle either. Between I know I have a purpose and I don't care if I have a purpose, sitting on the fence is no better. You were created with a specific purpose for your life.

> *I know your deeds, that you are neither cold nor hot. I wish you were either one or the other! So because you are lukewarm, neither cold nor hot, I am about to spit you out of my mouth.*
> **Revelation 3:15-16 (NIV)**

A little bit of world with a little bit of church does not please Him.

Future: time or period of time following *the moment.*

There will be many moments in your life. Moments that will bring both joy and sadness. Moments of tears and laughter. Remember to claim your future, your destiny, you must keep moving forward.

Philippians 3:13b (NLT) Forgetting the past and looking forward to what lies ahead.

Your future is moving all the time. Your heavenly Father hasn't stopped working on your behalf. Getting you to your destiny of purpose. Father God doesn't sleep or slumber.

Psalm 121:3b-4 (NLT) He who watches over you will not slumber. Indeed, He who watches over Israel will neither slumber nor sleep.

Relax and trust.
As a child of God, enter His rest.

Hebrews 4:3a (NIV) Now we who believe enter that rest.

Keep your eyes focused on your future. Father God has great plans to use you and your struggles, sorrows and troubles for His glory.

Fix your gaze on the One who knows your every concern, stress, your time of loneliness and tribulations.

For our light and momentary troubles are achieving for us an eternal glory that far outweighs them all. So we fix our eyes not on what is seen, but on what is unseen since

*what is seen is temporary, but what is unseen
is eternal.*

2 Corinthians 4: 17-18 (NIV)

He knows that it is just for a moment, and things we see will soon be gone. One day we'll be joined again by the very ones we long to see.So my friend, let joy fill your heart and soul. Let your spirit rejoice. We have a destiny. We are predestined. We are called.

Romans 8: 30 (RSV) And those whom He predestined He also called, and those whom He called He also justified;and those whom He justified He also glorified.

And most importantly, one day we will see *love* face-to-face. Jesus, our King of Kings and Lord of Lords. I do believe He'll be just as happy to see us as we are to see Him. Our bridegroom!

Isaiah 62:5b (NKJ) As the bridegroom rejoices over the bride, so shall your God rejoice over you.

Jesus our healer. Our restorer.

Psalm 23:3a (NKJ) He restores my soul.

We all have a destiny to *go* and *do.*

Let us fulfill our purpose and go home having run our race to the glory of God.

Now there are steps to take. He always takes us step by step. You already know a step or two.

Begin to take them. Move on them. To get started, write them down. The ones you already have and then continue to add to your list. God is waiting to take you on a journey only He can take you on.

Destiny- fulfilling it step by step.

_____

_____

_____

_____

_____

_____

_____

_____

_____

_____

_____

_____

_____

_____

_____

_____

CHAPTER 11
CONVERGENCE

Colossians 3: 14 (NLT) Over all these virtues put on love, which binds them all together in perfect unity

**Convergence- the act of coming together.
Moving toward unity.**

The Holy Spirit, when converged with our spirit in unity, has filled us with love for others.

It's in you to love, serve, and bless those around you. As you release grief and move into God's purpose, you will sense the flowing of love from you to others. Helping others to be reconciled to their Heavenly Father. Helping them finish strong in their purpose. Then we all will share in His glory.

Part of our purpose is to take as many friends, relatives, and co-workers to heaven with us!

So converge your spirit with the Holy Spirit and in love share your story, your loss and your joys with others.

Love.

That is the word that birthed you in this world. Father God's love for you, for His children, for a family.

We are family. God's family. Praise God! We will return to our family in heaven when we have finished our race, our purpose, and our destiny. So for the last journal entry, let's write a letter to the One who knew us before we were formed in our mother's womb.

Jeremiah 1:5a (NIV) Before I formed you in your mother's womb, I knew you. Before you were born I set you apart.

The One who formed us in our mother's womb.

Psalm 139:13 (NKJ) For you formed my inward parts; You covered me in my mother's womb.

Who loved us first?

1 John 4:19 (NKJ) We love Him because He first loved us.

Spend a moment or may I say moments listening, hearing, healing, getting understanding and wisdom. Heavenly Father wants you to have it. So you know your perfectly planned purpose for such a time as this. As you read through the Bible, notice how Father revealed to His children the purpose and plan He had for them. You are no different. He wants you to know. He wants you to finish your race with joy and hope.

Now share your heart with Him. Then sit quietly and

let Him share His heart with you. Let me add,He is always talking to us. We just need to listen. And let *Him* do most of the talking I might add.

James 1:19 (NKJ) So then my beloved, let every man be swift to *hear*, slow to *speak.*

To be able to hear Father God speak to you,you must know Him. And that is by receiving His love through His only begotten Son, Jesus. If you have invited Him into your heart and life, Hallelujah!

You are ready to sit, talk and listen.

If that is not you yet, let me help you right here and right now. He has been waiting for you. Longing for you. Seeking you.You just need to know He loves you, wants you and will accept you as His own this very minute. You will receive His peace through Jesus. You will no longer be separated from your creator, your heavenly Father. We were brought back to our Father through the blood and sacrifice of Jesus. So come with me into His presence and let's pray together and then visit with Father God.

Heavenly Father,

I long to get to know you and why I am here on this earth. So I decide right here and right now to believe that Jesus died for my sins and was raised from the dead to bring me back in relationship with you my Father in heaven. Forgive all my sins and make me right with You. I trust Jesus as my Lord and Savior from this day forward. Teach me, lead me and talk to me. Show me my purpose in You. Amen.

> *If you confess with your mouth the Lord Jesus and believe in your heart that God has raised Him from the dead, you will be saved. For with the heart, one believes unto right standing with God and with the mouth confession is made unto salvation.*

Romans 10:9-10 (NKJ)

Welcome to the family!

Now go spend time with your Father.

Hello Father, can we talk? and then I will listen

Chapter 11: Convergence

_____

_____

_____

_____

_____

_____

_____

_____

_____

_____

_____

_____

_____

_____

_____

_____

_____

_____

_____

EPILOGUE

Well, here we are. Thank you for letting me share my heart with you. This,I believe, is part of my purpose, my destiny, my future and my race. I hope to run it well. I hope I was able to share my journey and offer you comfort, encouragement and strength to continue to work out your salvation.

Philippians 2:12b (NLT) Work hard to show the results of your salvation.Obeying God with deep reverence and fear.

I have taken what Father God has given me and I am working it out in my life. The happiness I feel and the joy and fulfillment I feel in my heart, I would never have thought possible after seeing my daughter graduate to heaven.I can happily admit to now having a renewed peace and eagerness has filled my heart and soul. I wake up each morning with a prayer on my lips.

"Father, use me to help someone today who needs a word of encouragement, to be shown love, a hug or that I can pray for."

He does use me as I *allow* Him to.We must be willing. Be willing today my friend. You will experience so much excitement, joy, and a reason to go on. We must go on until we are called home and graduate to our new home with Father God and Jesus.Then, the reunion will be amazing! I'm looking forward to it.

Such joy.

John 16:22 (NIV) Now is your time of grief, but I will see you again and you will rejoice, and no one will take away your joy.

P.S.

Do let me know how your journey is going and if I can pray for you. Remember, none of us were meant to journey alone. We are part of the family of God. We are His body. His Temple.

Ephesians 2:21 (NLT) We are carefully joined together in Him, becoming a holy temple for the Lord.

Stay close to Him my brothers and sisters, because He is always near.